3 2783 00056 2119

Constructivism and Science Teaching

by
Alan Colburn

ISBN 0-87367-635-1
Copyright © 1998 by the Phi Delta Kappa Educational Foundation
Bloomington, Indiana

PRAIRIE STATE COLLEGE
LEARNING RESOURCES CENTER

This fastback is sponsored by the
Illinois State University Chapter of
Phi Delta Kappa International,
which made a generous contribution
toward publication costs.

Table of Contents

Introduction

Imagine yourself in an inservice workshop. The facilitator takes out a pie tin, birthday candles, matches, a container of water, and an empty glass flask. She sticks one candle to the middle of the pie tin and pours water around its base. Then she lights the candle.

With a smile on her face, she gingerly places the flask over the burning candle. Several things happen. When she calls for observations, participants recount seeing the candle go out, smoke filling the flask, water rising into the flask, and bubbles appearing in the pie tin just outside the flask.

Next the facilitator asks a more difficult question, "What do you think made the water rise?" Various participants offer different explanations. A popular reason is that the burning candle used up the oxygen in the flask. When the oxygen was gone, the candle went out and the water rose to take the place of the oxygen.

But some participants offer an alternative reason. The candle heated the air in the flask. When the air was heated, it expanded. The expanded air made the bubbles that the participants observed. After the candle went out and the air cooled, water came into the flask to fill the space once occupied by the heated air.

One participant offers yet another explanation. Hot air rises. The candle heated the air inside the flask, the heated air rose to the highest part of the flask, and water then came into the flask to take the place of the now-risen air.

The facilitator summarizes these three explanations on the chalkboard and challenges the participants to figure out the correct explanation. Participants are stymied. "Suppose," offers the facilitator, "that you lit three candles instead of one. What differences, if any, would you predict with each of the hypotheses?"

With the realization that the two more popular hypotheses predict different outcomes, participants eagerly begin experimenting on their own. Some simply repeat the facilitator's demonstration, watching what happens a little differently than they did the first time. Others try using more candles. Two participants try raising the level of the candle inside the flask. They explain that they think the candle will stay lit longer this way — no rising water to extinguish the flame — and that the water eventually will rise higher in the flask.

Throughout the activity, the facilitator watches the participants' actions. Now and then she helps with materials. More often, she will approach a group and ask them what they are doing or thinking at that point. She usually responds to their comments with another question or a suggestion. Sometimes she points out something that other participants are doing.

When the group reassembles, most people agree on a single explanation. The ensuing discussion centers on aspects of the activity that made the difference in helping participants change their minds and why lecture alone would not have had the same effect.

This is constructivism-based teaching in action, and it is as easy to manage with children as with adults.

In this fastback I outline the basics of constructivism in the context of teaching science. What does it take to apply constructivism-based teaching principles? Instructional practices include using open-ended, hands-on activities; cooperative learning; various questioning strategies; student journals; and a curriculum based on the learning cycle. I suggest how to make a gradual transition to such teaching for teachers who are unaccustomed to constructivist practices. I also discuss assessment and the place of lectures and textbooks in the constructivism-based classroom. Finally, I consider some of the varieties of constructivism-based teaching currently being discussed, such as radical constructivism and interactive constructivism.

What Is Constructivism?

When the Soviets launched Sputnik in the late 1950s, a public perception was fostered that U.S. science education was in crisis. In the years since, science education has not lacked for experts with various solutions to the crisis. Fads and buzzwords abound, mixing freely with curriculum and instructional practices that hold real promise for improving what and how students learn science. Constructivism connects disparate ideas and provides a way to winnow the wheat from the chaff.

Theory

Constructivism is a theory about the nature of reality and how people understand the world around them. (For a general view, see fastback 390 *Constructivist Teaching*, by John A. Zahorik.) Constructivists argue that humans make (construct) their own knowledge based on their experiences. This knowledge creates the individual worldview that each student brings into the classroom. What students hold to be "true" will be based on what mentally works for them, what makes sense within their conceptual framework — what von Glaserfeld (1992) refers to as *viability.*

According to constructivist theory, ideas are not held to be absolutely true or false. Instead, ideas explain and predict in ways that are better or worse than other ideas. In other words, some ideas are more viable than others. Rather than talking about what is "true" in science, for example, constructivists talk about what is generally agreed on by the majority of the scientific community.

From a constructivist viewpoint, science teaching involves helping students understand how and why some knowledge explains and predicts more accurately than other (prior) knowledge (or beliefs) by providing experiences and opportunities that encourage students to construct accurate knowledge. Because no student is void of knowledge (even though much that students believe about scientific matters may be limited), learning science involves replacing some ideas with others. Thus new knowledge is used to correct prior knowledge. This view contrasts with the notion that teachers are "givers" of knowledge. Rather, students must be "makers" of knowledge.

The demonstration I recounted in the Introduction illustrates this concept. Many participants were initially satisfied with the explanation that the water rose because used-up oxygen created a partial vacuum inside the flask. That explanation has a reasonable "feel" to it. But it is not the best response in this case. Scientists generally accept the explanation that begins with the candle heating air inside the flask. And the further investigation led the participants to arrive at this conclusion themselves through the process of experimentation and discussion.

New experience, therefore, encouraged the participants to construct new knowledge that replaced previously held beliefs. This is an important part of the constructivism-based teaching process in action.

Prior Knowledge, or "Intuitive Science"

It may be useful to look a bit more deeply into the process of constructing knowledge. Educators use a variety of terms when referring to the personal beliefs that students hold about a topic. Common terms include "preconceptions," "misconceptions," "personal theories," and "naive beliefs." Intuitive beliefs, or naive beliefs, are not stupid, but merely uninformed. To distinguish such beliefs from ideas that tend to be better accepted by the scientific community, I will use the terms "intuitive science" and "misconceptions" as synonyms throughout this fastback.

Scientific knowledge does not necessarily match students' "commonsense" understandings; the opposite is sometimes true, in fact (Dorsett 1988). Pinker (1997) points out that children, adults, and scientists all have to make sense of the world, are curious, and try to turn their observations into valid generalizations. But humans evolved to cope with their environment, not to publish scholarly science papers. As Pinker puts it, "Good science. . . was an unlikely selection pressure within illiterate foraging bands like our ancestors, and we should expect people's natural 'scientific' abilities to differ from the genuine article" (p. 306).

Several points are worth considering about intuitive science, the prior knowledge with which teachers must work when they begin to help students construct science knowledge. First, intuitive science makes sense to the believer. Second, intuitive understandings often are based on experience and are strongly held. Third, many students are unaware of the intuitive beliefs they hold until they begin to examine them.

Many intuitive beliefs reflect beliefs once commonly held to be true. Animal adaptation (for example, fur growth in cold climates) often is believed to be willful, a once commonly held view that had been expressed by Lamarck, an 18th century scientist. Many children and adults talk about the sun rising, crossing the sky, and falling, even though they are taught early on that it is the earth that moves, not the sun. The intuitive notion reflects a Ptolemaic view that was regarded as true for many centuries.

Challenging — and Changing — Intuitive Science

The duty of the science teacher in applying constructivist principles is to help students change their thinking.

This process begins with challenging intuitive beliefs, making the learner dissatisfied with his or her misconceptions. Piaget called this "cognitive disequilibration" or "cognitive dissonance." Intuitive views cannot be changed without this initial dissatisfaction.

But conceptual change requires more than dissatisfaction. A new, alternative belief also must be available.

The new understanding must explain and predict better than the previous one. In some ways, learning is like jumping from one ship to another in the middle of the ocean. People will not make the jump unless (a) something is drastically wrong with the old ship, and (b) the new ship is in better shape than the old one.

To continue this analogy, the larger the ships involved, the greater the distance one has to jump and the greater the danger to the jumper. Similarly, the larger and more important the beliefs are to the learner, the more the learner will resist the new knowledge. In the example that opened this fastback, giving up the idea that water rose in a flask because oxygen was used up represented a minor idea to many of the observers. Being little invested in the concept of the experiment, they were willing to change their thinking about it. Larger ideas, more central to a personal understanding of the world, are more resistant to change.

The ship jumping analogy brings in another aspect about conceptual change. Resistance to new knowledge often arises because individuals would prefer to repair the old ship, rather than jump to a new one. Similarly, learners generally try to accommodate their current beliefs, rather than discard them and adopt totally different beliefs.

Teaching in the Constructivism-Based Classroom

Conceptual change is the central goal of the constructivism-based classroom. Remember, though, that constructivist ideas are not appropriate or relevant to *all* educational goals. With this in mind, conceptual change happens when:

- The teacher and the students are aware of prior knowledge, misconceptions, and intuitive science;
- The students become dissatisfied with the limitations of their intuitive beliefs; and
- The students participate in activities that challenge prior knowledge and enable them to construct new understandings.

These things happen in classrooms where students actively study various scientific phenomena and communicate about beliefs and observations. Someone walking into a constructivism-based science classroom will see students moving about, engaging in experiments, making observations, asking and answering questions, keeping journals, and discussing their learning with the teacher and other students.

The following sections describe several of these key processes.

Inquiry-Based Teaching

Let us walk into Ms. Smith's science class. Ms. Smith is showing her students a simple pendulum. She asks, "If you want to make this pendulum swing back and forth 60 times in a minute, like the second hand on a watch, what might you do?" Students brainstorm ideas, and Ms. Smith writes them on the chalkboard.

Next, Ms. Smith distributes materials, and the students begin trying to figure out how to make a pendulum swing back and forth once each second. In a few moments it becomes obvious that the students are trying a variety of techniques. Some are manipulating the length of the pendulum; others are trying different weights on the end of the shaft.

Inquiry has two broad meanings, one associated with questioning and the other with investigation. In an inquiry-based classroom, such as Ms. Smith's, students will be engaged in both types of inquiry. I will say a bit more about questioning in a later section. For now I want to focus on investigation.

Ideally, in an inquiry-based classroom authentic investigation is common. Such investigation may be stimulated by a problem posed by the teacher — the pendulum question, for example — or by students' natural curiosity. Investigation may be designed to explain some observed phenomenon or to provide information on which to base a decision (Haury 1993). In any case such inquiry stimulates students to think not only about the topic under study but also about their prior beliefs.

Inquiry-based teaching often is described as "hands-on" or "process-oriented," but neither characterization adequately captures its essence. "Hands-on" science — merely having students manipulate science materials — does not ensure learning. And a process approach needs more explanation, such as that offered by the National Science Education Standards:

> The Standards call for more than "science as process," in which students learn such skills as observing, inferring, and experimenting. Inquiry is central to science learning. When engaging in inquiry, students describe objects and events, ask questions, construct explanations, test those explanations against current scientific knowledge, and communicate their ideas to others. They identify their assumptions, use critical and logical thinking, and consider alternative explanations. In this way, students actively develop their understanding of science by combining scientific knowledge with reasoning and thinking skills. (National Research Council 1996, p. 2)

I would illustrate these distinctions by referring again to the pendulum activity in Ms. Smith's classroom. A common "hands-on" science activity has students demonstrating that the length of a pendulum affects how long it takes for the pendulum to make a complete swing. The longer the pendulum, the more time required for the swing. Students might be shown how to build a pendulum of a given length. They might be given a picture of the desired lab setup. Students then would be asked to swing the pendulum a given number of times and to record the time required for a complete swing in a lab workbook. Next they might be asked to modify the setup, to make pendulums of other specified lengths, to repeat the procedure, and again to record the results. Finally, the students would be asked questions about how the length of the pendulum affects the time for a swing.

While this form of the activity is a step in the constructivist direction, it is not inquiry-based teaching. The activity I have described can be completed by the students with little real thought or effort. By contrast, Ms. Smith's approach puts more responsibility in the hands of the students. They must pose questions related to Ms. Smith's initial challenge, and they must come up with ways to test their intuitive science and correct or replace prior knowledge with new information from their investigation.

Cooperative Learning

Let us return to Ms. Smith's classroom and change the scenario slightly. After Ms. Smith poses her initial question, instead of writing the brainstormed ideas on the chalkboard, she asks the students to gather into pre-assigned groups. She then instructs each student to write down his or her ideas on a slip of paper. When they have done so, Ms. Smith asks them to share those ideas with the other members of their group. (Each student also places the written idea on a pile in the center of the table at which the group is sitting. This serves as an individual checkpoint for the teacher.)

Next the students work in pairs to experiment with pendulums. Each student in a pair takes turns manipulating the pendulum, timing the swings, and so on. They record their observations of each trial. At the end of the inquiry time, the two students compile their results, draw a conclusion, and sign their report to certify that they worked together and came to an agreement.

This scenario describes a way to layer cooperative learning onto the foundation of inquiry-based teaching. Cooperative learning is a broad term that encompasses many teaching methods that have in common the strategy of helping students learn together — with and from one another — in pairs and small groups. However, cooperative learning neither blunts the need nor lessens the effect of individual learning.

Theorists argue about the merits of cooperative learning. Some believe learning to be primarily a *social* enterprise, in which case cooperative learning should be a dominant feature of the classroom. Others argue that students need considerable *individual* time to process new learning, in which case cooperative learning should play a smaller role. But almost all agree that some amount of cooperative learning is valuable in a constructivism-based classroom.

Simply put, students of all ages often learn better when they have the chance to talk about the ideas they are studying with someone else. From a constructivist viewpoint, cooperative learning means that student talk is focused on the topic of study. Through this peer interaction, each student's ideas are challenged. Students are forced to explain their observations and conclusions and, in the process, to construct new knowledge.

A number of education theorists and practitioners have written about what is, or is not, cooperative learning. For my purposes, Kagan (1992) provides a lucid description of how cooperative learning differs from other types of group learning situations. In cooperative learning:

- Students are positively interdependent on one another. That is, they work in ways such that when one student benefits, all of the students in the group benefit.
- Students are held individually accountable for their learning. All students are expected ultimately to learn the same things. Students take tests individually, for example. Group grades are avoided (though some cooperative learning advocates believe group grades are acceptable under limited conditions).
- Students participate equally. No student dominates, no student sits passively while others learn.
- Students participate simultaneously. In cooperative learning situations, many students are active at the same time.

Questioning

Ms. Smith's students are working in cooperative groups on the pendulum swing inquiry. As they work, Ms. Smith circulates through the classroom, occasionally pausing to ask a question. She says to one pair of students, "What are you thinking at this point?" One answers, "We need to slow it down, so we're adding weight." Ms. Smith follows up the answer by asking what the student means by "slow it down." The student responds that the pendulum is now swinging 63 times in a minute.

"So," says Ms. Smith, "you're predicting that as the weight on the end of the pendulum is increased, the number of swings per minute will decrease."

"Right!" the student replies.

"Why do you think that might be?" asks Ms. Smith. "Try it out, think about it, and then let me know."

Questions are how teachers probe student thinking and find out the intuitive science students bring to class. They also serve to challenge that thinking and to pose new ideas that will help students toward conceptual change.

From a constructivist perspective the most effective questions start with students' own ideas or misconceptions and then stimulate students to examine and extend their thinking. The questions need not be complex to achieve this end. For example, during an activity the teacher might ask, "What are you doing?" or, as Ms. Smith asked, "What are you thinking at this point?" Such questions encourage students to articulate. One-word answers are impossible. And such questions also are non-threatening. The answer can be neither right nor wrong.

Related to the "what are you doing" and "why are you doing it" questions are those that ask students to look ahead: "What do you predict will happen?" This type of question urges students to think about — and put into words — what might happen as a result of various manipulations. Then they can test their prediction and draw a conclusion.

These forms of questions require teachers to have patience — that is, to allow sufficient "wait time" for students to mentally compose their answers. More than 20 years ago Rowe (1974) pointed out the benefits of extending wait time, which include increasing the number and length of student responses and helping students to become more confident in expressing themselves. More elaborate responses also help the teacher by providing more information that can be paraphrased or extended to make a point.

Studies have shown that many teachers wait a second or less for students to respond. Extending the wait time to five seconds may seem like an eternity, but wonderful things begin to happen when students have time to think and form their responses. Extended wait time also increases the number of students who desire to respond and so helps students hear a greater diversity of ideas. Discussion is enlarged and enriched in this way.

(For a useful resource, see Kenneth R. Chuska's *Improving Classroom Questions*, published by the Phi Delta Kappa Educational Foundation in 1995.)

Student Journals

After Ms. Smith's students have experimented with the pendulums for a few minutes, Ms. Smith goes to the chalkboard and writes, "I used to think ___, but now I think ___." Then she turns to the class and says, "Think about the activity you have been doing and complete this sentence by filling in the blanks. If you've changed your thinking about several things, then write about all the ideas that changed. Then go on to say why you changed your mind."

Many teachers use student journals as a way for students to record their procedures, results, and conclusions in a science activity. This is a valuable use, albeit limited. Constructivism-based teaching takes the student journal activity a few steps further.

In constructivism-based teaching, student journals become vehicles for conceptual change, in addition to assessments of student work. Shepardson and Britsch (1997) describe four times during the course of a science activity when journals can increase student learning: before, during, and after an investigation and, on reflection, to create a basis for more formal communication. This last use merits further explanation.

Most journal writing — the before, during, and after pieces — are intended to be "of the moment." Journals are a place for students to react to new ideas, explore their thoughts, make mistakes, correct them, and work through problems. The fourth time for journal writing is later, when the student can reflect on what happened during the science activity. This fourth writing gives the student an opportunity to read and correct the spur-of-the-moment writing. In so doing, the student also creates a basis, perhaps even a first draft, for a more formal paper.

Writing also helps students to understand and to articulate their thoughts. Writing ideas in a student journal gives the student a basis for discussion, and listening to other student responses also helps students reflect on their own notions.

Finally, another use of student journals is assessment. For this purpose, the journal must be examined over time. During the course of a few weeks or a semester, a student will set down misconceptions, try out new ideas, and eventually acquire new understandings. None of this process will be evident at a glance. Change takes time. Thus a journal kept over an extended period likely will provide a fuller picture of a student's ideas and how they change.

Two basic guidelines are important. First, teachers should clearly explain how journals are to be used. Students need to understand that keeping a journal is an integral part of learning. Second, teachers can reduce the burden of reading every word in every student's journal by:

- Asking the student to select portions of the journal to which the teacher will respond.
- Choosing portions of each student's journal to read and comment on.
- Interacting with students as they write either individually or in groups.
- Conducting mini-conferences with students during which they share their journals.
- Developing peer journal-sharing strategies.

Demonstration Teaching

In the pendulum experiment Ms. Smith merely showed the class a pendulum and posed a question to be investigated. But let us suppose she is teaching another activity. Ms. Smith might want to start her students thinking about why things sink or float by doing a simple demonstration.

"Will this orange sink or float?" Ms. Smith notes how many students predict each alternative.

Ms. Smith then places an orange into a tank of water. The students can see that the orange is floating. "What do you think will happen if I peel this orange," asks Ms. Smith. "Will it still float, or will it sink?"

The students offer both possibilities again, and Ms. Smith tallies their opinions on the chalkboard. Then she peels the orange and places it in the water again. It sinks.

"What happened?" asks Ms. Smith. "Why do you think the peeled orange sank?"

Demonstrations such as this one are ideal for activating student thinking using observation and questioning. Students can relate intuitive beliefs to actual events in the demonstration, and then they can initiate variations of the demonstration during lab time or cooperative learning situations.

Ms. Smith uses prediction in two instances, which starts students thinking about floating and sinking. But it is the crucial "why do you think" questions that are the focus of this learning activity. Why do you think the unpeeled orange floats? Why do you think the peeled orange sinks? What properties of the peel do you think are important? These questions will lead to discussion that draws out students' intuitive beliefs, helps them form hypotheses, and sets them on the road of experimentation that will follow.

Using the Learning Cycle

Consider a high school biology class in which Mr. Jones' students are learning about common fungi, such as molds and mildew. First, the students examine moldy bread and discuss what they know about mold and mildew in their homes. How can mold be controlled? What factors favor its growth? What inhibits the growth of mold?

After this discussion, Mr. Jones shows the students how to cultivate a sample of mold to use in the classroom. When the mold has developed, the students experiment to see what factors encourage or inhibit its growth.

Next, the students discuss the results of their experiments. Mr. Jones facilitates the discussion, which he then extends into an explanation of mold structure. He also relates the molds created by the students to other types of fungi.

Finally, the students experiment with yeast, a different fungus. This activity proceeds in much the same way: Students grow yeast and then experiment to discover growth stimulants and inhibitors.

This scenario is an example of the learning cycle, an instructional strategy that divides teaching and learning into three to five major phases. This strategy is known by various names; however, the concept generally is credited to the SCIS (Science Curriculum Improvement Study) program, one of the 1960s "alphabet soup" of elementary science programs. The "5E Model," popularized by the Biological Science Curriculum Study (1993), is another example of a learning cycle approach to instruction. The five E's of the model are *engage*, *explore*, *explain*, *expand*, and *evaluate*.

- *Engage* refers to both motivation — engaging students' attention and interest — and activation of prior knowledge. Examining the moldy bread was Mr. Jones' way of getting students to think about mold and to discuss what they knew, whether correct or not, about mold. During the discussion the teacher can begin to learn some student beliefs about mold and mold growth.
- *Explore* refers to the activity phase: growing the mold sample and experimenting to see what factors encourage or inhibit mold growth.
- *Explain* is the phase in which students reflect on their experiments (or experiences) and articulate the new understandings they have constructed. This also is the phase in which the teacher most directly introduces the students to new ideas.
- *Expand* refers to applying new understandings in new situations. For example, replicating the mold experiments using yeast was a way to expand and solidify new understandings.
- *Evaluate* refers to the end of the cycle, when the students can reflect on the total experience and when

the teacher can assess the amount and sophistication of student learning.

But a learning cycle also means that some phases may serve to begin the cycle again. For example, an explanation or an expansion might provide the impetus for a new cycle of engagement, exploration, and so on. Evaluation, therefore, may be seen as formative — hence a stepping stone to a new cycle — or summative.

The undergirding philosophy of a learning cycle is, first, that learning will be more effective if the students have relevant, concrete experiences with content before being formally introduced to it. Advocates of the learning-cycle approach stress that direct experience should precede verbal instruction so that students can relate abstract verbal content to the concrete experience.

Second, students will learn more effectively if they engage in activities that are open-ended or inquiry-based. Students need to *do* something, to use ideas in active ways and to experiment, observe, and draw conclusions. The teacher and students work together in ways that help the teacher better understand students' misconceptions about the topics being studied.

Finally, once students have been formally introduced to new ideas, they should mentally *do* something with the ideas in a new context. This gives some students a chance to solidify their learning or, in other cases, to begin the process of conceptual change.

Assessment

Ms. Smith begins to assess students' prior knowledge — and to help students assess their own knowledge —

as soon as she completes the orange demonstration. First, she asks, "What did you just see?" This raises observation and speculations.

Next, Ms. Smith asks the students to make a KWL chart. This chart has three columns labeled:

K = What do I *know*?
W = What do I *want* to know?
L = What did I *learn*?

Using the KWL chart, the students individually record their prior knowledge — the K column — and then pose questions that they want to answer — the W column — during the experimentation/exploration phase. As they answer those questions, they will fill in the L column. In many cases the same topic appears across all three columns, but sometimes there are more questions in the W column or unanticipated new understandings in the L column. One-to-one correspondence is not essential.

Constructivism-based teaching proceeds from "authentic" assessment. That is, the teacher assesses student knowledge and learning based on experiences that are as realistic as possible. Students solve real problems, not theoretical ones. Traditional, pencil-and-paper tests are rarely used in this type of assessment. At the same time, there are pencil-and-paper products. The KWL chart is an example. And such a chart may be freestanding or part of a student journal.

Another important concept in constructivism-based teaching is the use of ongoing assessment. Teachers gather assessment information — through observation, questioning, viewing students' journals or KWL charts, and so on — during instruction. By so doing they can determine when to intervene if students' experiments become unproductive or if students are jumping to incorrect conclusions, for example.

Summative assessment also can be informed by student journals, direct observation, questioning and discussion, and, when necessary, by formal tests.

Making the Transition to Constructivism- Based Teaching

Typically, junior high, middle school, and high school science teachers combine direct instruction (lecture), cookbook-type laboratory activities, and independent student seatwork. Moving from this traditional style of instruction to a more constructivist style can be difficult for teachers and students. After all, many students also are accustomed to traditional classroom practices. Constructivism-based teaching requires teachers and students to take on new roles. However, a gradual transition can ease the way.

In the next few sections I suggest constructivism-based teaching strategies and how to implement them in making the transition from traditional teaching to constructivism-based practice.

Do the Lab First

Lecture-then-lab is a time-honored sequence in the traditional science class. One transitional adaptation is to reverse the sequence. Teachers can teach the lab activity, making only minimal changes to the activity to ensure that students can complete it successfully.

By moving to the lab before the lecture, teachers enhance the spirit of inquiry. Students will not know what is supposed to happen and so can make that discovery on their own. Thus the lab becomes a way of activating and testing prior knowledge as it actively engages the student in the new learning topic.

Similar strategies have been advocated by such groups as Chem Study and PSSC Physics, which means that lab-then-lecture is not an unknown strategy. It has been classroom tested and found to be successful.

Discuss the Lab, Then Lecture

This step naturally follows the previous one. Teachers need to give students a structured opportunity to process the information generated by the lab activity. Rather than launching into a lecture right after the lab, teachers can lead a discussion about students' inquiry experiences:

"What were you investigating?"

"Why do you think the lab procedure was set up in this particular way?"

"What were your results?"

"What interpretations do you think can be made about the meaning of the data?"

"What have you learned from doing the activity?"

The goals of this discussion are to help students think about 1) what they believed before the activity, 2) what they discovered during the activity, and 3) what questions were raised by the activity. The lecture that follows this discussion should answer those questions — or help students figure out how to answer them.

Remove the Data Table

The first two strategies simply change how a typical cookbook lab activity might be treated nontraditionally. A further transitional strategy is adapting the lab activity itself.

One initial step can be to remove the data table from the lab manual. Most lab manuals provide a formatted data table so that students simply have to fill in the blanks. To push up the inquiry value of the activity, teachers can ask students to determine how data should be displayed, keeping in mind that the data display should be intelligible to someone who has not performed the activity.

This is an effective transitional strategy because the teacher still chooses the activity and the procedures. The only real change — but an important one — is that students must develop their own data tables. But this is important because it forces students to think critically — and possibly creatively. Removing the formatted data table also means that there may be more than one way to display the data correctly.

Change the Test

Adapting the lab activity can be complemented by adapting the test for a given unit of study. Tests and quizzes communicate to students what the teacher believes is important. Therefore, the more reflective a test is, the more likely that students will value reflection on what they have learned and new knowledge they have constructed.

Ideally, any test should help students explore, consider, and decide matters for themselves. One strategy can be to ask students to interpret raw data from an activity similar to one they have done in the laboratory. Another approach is to ask students to predict an outcome or a change that might result from changing a part of the lab procedure.

Use More Questioning

This transitional strategy focuses on less lecture and more discussion. Adept teacher questioning can lead students to think for themselves and will encourage productive discussion in all settings, whether the students are working in small groups or engaged in a whole-class discussion.

The most productive questions are open-ended and nonthreatening. "What do you mean by ___?" and "Why do you think ___?" are good ways to start. Such questioning helps students clarify their thinking and raises points of inquiry. Questioning actively engages students, whereas a lecture usually results only in passive student engagement.

Help Students Invent Procedures

As students become more comfortable taking on active inquiry roles, they can be asked to take on greater responsibility for their own learning. One strategy is to help students begin to develop, or "invent," their lab procedures. This is a good next step after omitting data tables, for example.

The teacher might provide only the question to be answered, leaving the students to determine the materials to be used, the experiments to be conducted, and how to arrange the data they generate. At first, it may be necessary for the teacher to demonstrate a lab setup or to provide hints about essential materials; but over time students will acquire the necessary inquiry skills to carry on independently.

Persist in the Transition

These suggested transitional strategies are starting points. As teachers and students change roles, it is important to persist in the transition. Constructivism-based teaching is stimulating and exciting, but making the transition from traditional practice to constructivism-based practice also can be daunting. Taking the small steps suggested previously can help both teachers and students make a smooth, gradual transition that will profit everyone and enhance student learning.

A Final Caveat

Constructivism means different things to different people. While constructivism-based teaching, in principle, has gained wide acceptance, constructivism itself can be open to interpretation.

The basic tenets of constructivism are easy to state: Learning is an active process. Learners construct new knowledge by relating new experiences and new understanding to prior knowledge. But beyond this simple definition there lie many variations.

For example, Good and colleagues (1993) list 15 adjectives that educators and others use when they describe constructivism:

contextual
dialectical
empirical
humanistic
information-processing
methodological
moderate
Piagetian

postepistemological
pragmatic
radical
rational
realist
social
socio-historical

Discussing the nuances of these descriptors is beyond the scope of this fastback. Suffice it to say that subtleties of interpretation color how constructivism-based practice is shaped *in theory*, though individual interpretation in the classroom likely can be viewed as more important on a day-to-day basis for teachers and students.

A turning point for interpretation is one's understanding of the nature of reality. To some, an objective reality exists, and humans have the ability to describe and understand that reality.

To others, things such as the senses limit, or affect, one's ability to understand the world. Humans view reality through the lens of personal knowledge and experience. To use the classic example, were a fish able to communicate, it would convey a very different understanding of water than would a human.

Ideas about reality affect how teachers and students construct meaning and thus affect constructivism-based practice in the classroom. Perhaps even more important, however, is how teachers view student knowledge. A deficit view of student knowledge — intuitive science — is that such misconceptions are "wrong." Constructivism posits the opposite view, that intuitive science represents students' attempts to construct meaning based on limited experience. Therefore the role of the teacher is to move students to construct new knowledge that corrects or changes limited prior knowledge.

In developing this fastback on constructivism-based practice in the science classroom, I have attempted to articulate a view of constructivism-based teaching that most of my colleagues will find agreeable. Few would argue with at least some use of open-ended hands-on activities, cooperative learning, less lecture and more questioning, students journals, and so on. But interpretations of constructivism vary somewhat. My goal has been not to advocate any particular set of practices but to suggest some successful strategies for those educators who desire to make the transition from traditional practice to constructivism-based practice that can energize science teaching and learning.

Resources

Biological Sciences Curriculum Study. *Developing Biological Literacy*. Dubuque: Kendall/Hunt, 1993.

Colburn, A., and Clough, M. "Implementing the Learning Cycle." *The Science Teacher* 64, no. 5 (1997): 30-33.

Dorsett, B. "Knowledge of Science Vital for All Children." *Education Week*, 23 November 1988, pp. 28, 22.

Finley, F. "Why Students Have Trouble Learning from Science Texts." In *Science Learning: Processes and Applications*, edited by C.M. Santa and D.E. Alverman. Newark, Del.: International Reading Association, 1991.

Good, R.; Wandersee, J.; and St. Julien, R. "Cautionary Notes on the Appeal of the New "Ism" (Constructivism) in Science Education." In *The Practice of Constructivism in Science Education*, edited by K. Tobin. Washington, D.C.: AAAS Press, 1993.

Haury, D. "Teaching Science Through Inquiry." *ERIC/CSMEE Digest* (March 1993): 4

Kagan, S. *Cooperative Learning*. San Juan Capistrano, Calif.: Resources for Teachers, 1992.

Leonard, W. "A Recipe for Uncookbooking Laboratory Investigations." *Journal of College Science Teaching* 21, no. 2 (1991): 84-87.

Murphy, N. "Helping Preservice Teachers Master Authentic Assessment for the Learning Cycle Model." In *Behind the*

Methods Class Door: Educating Elementary and Middle School Science Teachers, edited by L.E. Schafer. Columbus, Ohio: ERIC Clearinghouse for Science, Mathematics, and Environmental Education, 1994.

National Research Council. *National Science Education Standards*. Washington, D.C.: National Academy Press, 1996.

Pinker, S. *How the Mind Works*. New York: W.W. Norton, 1997.

Roth, K. "Reading Science Texts for Conceptual Change." In *Science Learning: Processes and Applications*, edited by C.M. Santa and D.E. Alverman. Newark, Del.: International Reading Association, 1991.

Rowe, M. "Science, Silence and Sanctions." *Science and Children* 34, no. 1 (1996): 35-37.

Rowe, M. "Wait-Time and Rewards as Instructional Variables." *Journal of Research in Science Teaching* 11, no. 2 (1974): 81-94.

Shepardson, D.P., and Britsch, S.J. "Children's Science Journals: Tools for Teaching, Learning, and Assessing." *Science and Children* 34, no. 5 (1997): 13-17, 46-47.

von Glaserfeld, E. "Questions and Answers About Radical Constructivism." In *Relevant Research*, vol. 2, edited by M. Pearsall. Washington, D.C.: National Science Teachers Association, 1992.

PRAIRIE STATE COLLEGE
LEARNING RESOURCES CENTER

3 2783 00056 2119